ENERGY SECTOR STANDARD OF THE PEOPLE' S REPUBLIC OF CHINA

中华人民共和国能源行业标准

Specification for Preparation of Special Geological Report on Impoundment-Affected Area for Hydropower Projects

水电工程水库影响区地质专题报告编制规程

NB/T 10129-2019

Chief Development Department: China Renewable Energy Engineering Institute

Approval Department: National Energy Administration of the People's Republic of China

Implementation Date: October 1, 2019

China Water & Power Press

中国水利水电出版社

Beijing 2024

图书在版编目（CIP）数据

水电工程水库影响区地质专题报告编制规程 : NB/T 10129-2019 = Specification for Preparation of Special Geological Report on Impoundment-Affected Area for Hydropower Projects（NB/T 10129-2019） : 英文 / 国家能源局发布. -- 北京 : 中国水利水电出版社, 2024. 9. -- ISBN 978-7-5226-2738-0

Ⅰ. P642-65

中国国家版本馆CIP数据核字第2024QX6751号

ENERGY SECTOR STANDARD
OF THE PEOPLE'S REPUBLIC OF CHINA
中华人民共和国能源行业标准

Specification for Preparation of Special Geological Report on Impoundment-Affected Area for Hydropower Projects

水电工程水库影响区地质专题报告编制规程

NB/T 10129-2019

（英文版）

Issued by National Energy Administration of the People's Republic of China
国家能源局　发布
Translation organized by China Renewable Energy Engineering Institute
水电水利规划设计总院　组织翻译
Published by China Water & Power Press
中国水利水电出版社　出版发行
Tel: (+ 86 10) 68545888　68545874
sales@mwr.gov.cn
Account name: China Water & Power Press
Address: No.1, Yuyuantan Nanlu, Haidian District, Beijing 100038, China
http: //www.waterpub.com.cn
中国水利水电出版社微机排版中心　排版
北京中献拓方科技发展有限公司　印刷
184mm×260mm　16 开本　2.75 印张　87 千字
2024 年 9 月第 1 版　2024 年 9 月第 1 次印刷
Price（定价）: ¥ 450.00

Introduction

This English version is one of China's energy sector standard series in English. Its translation was organized by China Renewable Energy Engineering Institute authorized by National Energy Administration of the People's Republic of China in compliance with relevant procedures and stipulations. This English version was issued by National Energy Administration of the People's Republic of China in Announcement [2023] No. 8 dated December 28, 2023.

This version was translated from the Chinese Standard NB/T 10129-2019, *Specification for Preparation of Special Geological Report on Impoundment-Affected Area for Hydropower Projects*, published by China Water & Power Press. The copyright is reserved by National Energy Administration of the People's Republic of China. In the event of any discrepancy in the implementation, the Chinese version shall prevail.

Many thanks go to the staff from the relevant standard development organizations and those who have provided generous assistance in the translation and review process.

For further improvement of the English version, any comments and suggestions are welcome and should be addressed to:

China Renewable Energy Engineering Institute
No. 2 Beixiaojie, Liupukang, Xicheng District, Beijing 100120, China
Website: www.creei.cn

Translating organizations:

POWERCHINA Chengdu Engineering Corporation Limited

China Renewable Energy Engineering Institute

Translating staff:

ZHOU Yifei   ZHANG Yixi   ZHAO Cheng   ZHANG Tao

HOU Hongying   CHEN Weidong   HU Jinshan   PENG Shixiong

Review panel members:

LIU Xiaofen   POWERCHINA Zhongnan Engineering Corporation Limited

QIAO Peng   POWERCHINA Northwest Engineering Corporation Limited

LI Zhongjie   POWERCHINA Northwest Engineering Corporation

Limited

QI Wen	POWERCHINA Beijing Engineering Corporation Limited
JIA Haibo	POWERCHINA Kunming Engineering Corporation Limited
PENG Peng	POWERCHINA Huadong Engineering Corporation Limited
YAN Wenjun	Army Academy of Armored Forces, PLA
WANG Shouyu	China Renewable Energy Engineering Institute

National Energy Administration of the People's Republic of China

翻译出版说明

本译本为国家能源局委托水电水利规划设计总院按照有关程序和规定，统一组织翻译的能源行业标准英文版系列译本之一。2023 年 12 月 28 日，国家能源局以 2023 年第 8 号公告予以公布。

本译本是根据中国水利水电出版社出版的《水电工程水库影响区地质专题报告编制规程》NB/T 10129—2019 翻译的，著作权归国家能源局所有。在使用过程中，如出现异议，以中文版为准。

本译本在翻译和审核过程中，本标准编制单位及编制组有关成员给予了积极协助。

为不断提高本译本的质量，欢迎使用者提出意见和建议，并反馈给水电水利规划设计总院。

地址：北京市西城区六铺炕北小街 2 号
邮编：100120
网址：www.creei.cn

本译本翻译单位：中国电建集团成都勘测设计研究院有限公司
水电水利规划设计总院

本译本翻译人员：周逸飞　张一希　赵　程　张　涛
侯红英　陈卫东　胡金山　彭仕雄

本译本审核人员：

刘小芬　中国电建集团中南勘测设计研究院有限公司
乔　鹏　中国电建集团西北勘测设计研究院有限公司
李仲杰　中国电建集团西北勘测设计研究院有限公司
齐　文　中国电建集团北京勘测设计研究院有限公司
贾海波　中国电建集团昆明勘测设计研究院有限公司
彭　鹏　中国电建集团华东勘测设计研究院有限公司
闫文军　中国人民解放军陆军装甲兵学院
王寿宇　水电水利规划设计总院

国家能源局

Announcement of National Energy Administration of the People's Republic of China [2019] No. 4

National Energy Administration of the People's Republic of China has approved and issued 297 sector standards such as *Code for Electrical Design of Photovoltaic Power Projects*, including 105 energy standards (NB), 168 electric power standards (DL), and 24 petrochemical standards (NB/SH).

Attachment: Directory of Sector Standards

National Energy Administration of the People's Republic of China

June 4, 2019

Attachment:

Directory of Sector Standards

Serial number	Standard No.	Title	Replaced standard No.	Adopted international standard No.	Approval date	Implementation date
…						
2	NB/T 10129-2019	Specification for Preparation of Special Geological Report on Impoundment-Affected Area for Hydropower Projects			2019-06-04	2019-10-01
…						

Foreword

According to the requirements of Document GNKJ [2014] No. 298 issued by National Energy Administration of the People's Republic of China, "Notice on Releasing the Development and Revision Plan of the First Batch of Energy Sector Standards in 2014", and after extensive investigation and research, summarization of practical experience, and wide solicitation of opinions, the drafting group has prepared this specification.

The main technical contents of this specification include: overview, regional geology and seismicity, basic geological conditions of reservoir, engineering geology segmentation and assessment of reservoir bank, geological principles for defining impoundment-affected areas, defining reservoir bank landslide-affected area, defining reservoir bank deformation-affected area, defining reservoir bank collapse-affected area, defining reservoir immersion-affected area, defining reservoir karst waterlogging-affected area, defining reservoir mined-out deformation-affected area, geological results for defining impoundment-affected area, and conclusions and suggestions.

National Energy Administration of the People's Republic of China is in charge of the administration of this specification. China Renewable Energy Engineering Institute has proposed this specification and is responsible for its routine management. Energy Sector Standardization Technical Committee on Hydropower Investigation and Design is responsible for the explanation of specific technical contents. Comments or suggestions in the implementation of this specification should be addressed to:

China Renewable Energy Engineering Institute
No. 2 Beixiaojie, Liupukang, Xicheng District, Beijing 100120, China

Chief development organizations:

China Renewable Energy Engineering Institute

POWERCHINA Chengdu Engineering Corporation Limited

Participating development organization:

China Three Gorges Corporation Ltd.

Chief drafting staff:

CHEN Weidong	PENG Shixiong	WANG Huiming	YUAN Jianxin
WANG Kui	MA Xingdong	XIE Jianming	ZHANG Shishu
WANG Xiaolan	HOU Hongying	YIN Xianjun	WANG Ping

Review panel members:

YANG Jian　LI Wengang　ZHU Jianye　ZHANG Yijun

MI Yingzhong　HUANG Minqi　LI Xuezheng　YANG Yicai

MENG Fanfan　ZHOU Zhifang　WANG Huiming　WU Yongfeng

WEI Zhenxin　YE Zhiping　CHEN Wenhua　ZHANG Sihe

LI Shisheng

Contents

1 General Provisions

1.0.1 This specification is formulated with a view to standardizing the preparation of special geological report on impoundment-affected area for hydropower projects.

1.0.2 This specification is applicable to the preparation of special geological report on impoundment-affected area for hydropower projects.

1.0.3 The preparation of special geological report on impoundment-affected area for small-sized hydropower projects may be simplified on the basis of this specification.

1.0.4 In addition to this specification, the preparation of special geological report on impoundment-affected area for hydropower projects shall comply with other current relevant standards of China.

2 Terms

2.0.1 reservoir bank landslide

impoundment-induced slide of rocks and soils in a reservoir bank along the sliding surface or belt

2.0.2 reservoir bank deformation

deformation of reservoir bank rocks and soils aggravated or caused by impoundment

2.0.3 reservoir bank collapse

collapse and slide of bank slopes of the reservoir under the action of water immersion, water level fluctuation, and wave erosion

2.0.4 reservoir immersion

salinization and swamping of soils, subsidence or damage of structure foundations, and water-filling or water inrush increase in underground works and mines, which are attributable to groundwater level rise induced by impoundment

2.0.5 reservoir karst waterlogging

impoundment-induced backward flow and flood detention in karst valleys or depressions around the reservoir

2.0.6 reservoir mined-out deformation

impoundment-induced surface deformation and caving-in of original goaf

3 Basic Requirements

3.0.1 At the stage of feasibility study of a hydropower project, geological work should be conducted to identify the range of impoundment-affected area, and the special geological report on the impoundment-affected area shall be prepared. After impoundment, geological work shall be carried out to identify new impoundment-affected areas according to the deformation and failure of reservoir bank, and a special geological report on new impoundment-affected areas shall be prepared.

3.0.2 The geological work shall ascertain the geological conditions of the reservoir area; investigate the hazard of bank slope deformation and its degree, and the importance of the affected objects; analyze and evaluate the engineering geological problems induced by impoundment, including the reservoir bank landslide, reservoir bank deformation, reservoir bank collapse, reservoir immersion, reservoir karst waterlogging, and reservoir mined-out deformation; and a special geological report on the impoundment-affected area shall be prepared.

3.0.3 Scoping of the impoundment-affected area for a hydropower project shall comply with the current sector standard DL/T 5376, *Specification of Scoping for Land Acquisition of Hydroelectric Projects*.

3.0.4 Suggestions on treatment, monitoring and patrol inspection for the impoundment-affected area should be put forward.

3.0.5 Special research and demonstration should be carried out for the impoundment-affected areas with a wide influence range, major hazard, important affected objects and high hazard.

3.0.6 The statistics of geological results for defining impoundment-affected areas should comply with Appendix A of this specification.

3.0.7 The special geological report on impoundment-affected area should include the main text, attachments, and drawings. The contents of special geological report on defining impoundment-affected areas should comply with Appendix B of this specification.

4 Overview

4.0.1 The overview shall briefly describe the geographic location, administrative division, physical and geographical conditions and transport conditions of the project.

4.0.2 The overview shall briefly describe the main structures, economic indicators, and relevant characteristics of the project; the main characteristic water levels of the reservoir, including normal pool level, minimum operating level, and limit level for flood control; the backwater calculation results, water level variation, principles and modes of reservoir operation.

4.0.3 The overview shall briefly describe the main basis, work process, main conclusions and workload of the geological investigation in the reservoir area.

5 Regional Geology and Seismicity

5.0.1 The regional geological conditions and the evaluation conclusions on regional tectonic stability shall be briefly described for the river or the river reach where the project is located.

5.0.2 The conclusion on seismic safety evaluation of the project site and the seismic ground motion parameters determination shall be briefly described.

6 Basic Geological Conditions of Reservoir

6.1 Topography and Geomorphy

6.1.1 For the reservoir area, the morphological characteristics, genetic type, adjacent valleys, and distribution and scale of oxbows, and geomorphic evolution of valleys shall be briefly described.

6.1.2 For the reservoir with a long shoreline, the types of topography and geomorphy in the river valley shall be briefly described by section.

6.2 Stratigraphy and Lithology

6.2.1 The distribution, genetic type and composition of overburden in the reservoir area shall be briefly described.

6.2.2 The type, lithology, distribution and combination characteristics of bedrock in the reservoir area shall be briefly described.

6.3 Geological Structure

6.3.1 The structural units and basic structural pattern of the reservoir area shall be briefly described.

6.3.2 The development pattern and characteristics of folds, faults and major joints and fissures in the reservoir area shall be briefly described.

6.4 Geophysical Phenomena

6.4.1 The development and distribution of physical geological phenomena such as landslide, collapse, deformation, and debris flow in the reservoir area shall be briefly described.

6.4.2 For karst areas, the karst collapse and accumulation shall be briefly described.

6.5 Hydrogeological Conditions

6.5.1 The types, burial, recharge, runoff and discharge conditions of groundwater, permeability of rocks and soils in the reservoir area and reservoir closure conditions shall be briefly described.

6.5.2 For karst areas, the characteristics of karst development and hydrogeological conditions shall be briefly described.

7 Engineering Geology Segmentation and Assessment of Reservoir Bank

7.0.1 The engineering geological segmentation of the reservoir bank shall be based on engineering geological mapping, exploration, and test results analysis.

7.0.2 The engineering geological segmentation of a reservoir bank shall consider the valley landform, geological structure types of the valley and reservoir bank, geological structure characteristics, rock and soil composition and properties, adverse geological processes and phenomena, hydrogeological conditions, reservoir bank stability, etc.

7.0.3 The engineering geological segmentation results of the reservoir bank shall be briefly described.

7.0.4 For the engineering geological assessment of the reservoir bank, the deformation and failure characteristics and stability status of each segment shall be analyzed.

8 Geological Principles for Defining Impoundment-Affected Areas

8.1 Basic Principles for Defining Impoundment-Affected Areas

8.1.1 The impoundment-affected areas may, according to geological factors, be categorized into reservoir bank landslide-affected area, reservoir bank deformation-affected area, reservoir bank collapse-affected area, reservoir immersion-affected area, reservoir karst waterlogging-affected area, and reservoir mined-out deformation-affected area.

8.1.2 The starting elevation for impoundment-affected area identification should be the envelope of normal pool level plus backwater height of corresponding flood frequency plus safety margin.

8.1.3 Impoundment-affected area shall be determined comprehensively according to the hazard of the reservoir bank slope, as well as the importance of affected objects and hazard degree.

8.1.4 The areas with major hazard, important objects affected and high hazard shall be regarded as impoundment-affected area.

8.1.5 The hazard of reservoir bank landslides and deformation shall be classified according to the stability coefficient, stability state and deformation and failure characteristics, and shall be in accordance with Table 8.1.5.

Table 8.1.5 Risk classification of reservoir bank landslides and deformation

Feature	Hazard			
	Major		Minor	
Stability coefficient K	$K < 1.00$	$1.00 \leq K < 1.05$	$1.05 \leq K < 1.15$	$K \geq 1.15$
Stable state	Unstable	Less stable	Basically stable	Stable
Deformation and failure characteristics	Creep to initial slip. The circumference of the sliding body is formed, and the sliding body goes off the sliding bed, with accumulated deformation of tens of centimeters	Creep, local cracks in the slope, accumulated deformation of tens of millimeters	Creep to no deformation, occasionally small cracks in the slope, accumulated deformation of millimeters	No deformation

8.1.6 For the reservoir bank slopes within the predicted bank collapse scope, the reservoir immersion and reservoir karst waterlogging areas with predicted elevations equal to or greater than the elevations of affected objects, and the reservoir mined-out deformation area with surface deformation, their hazards shall be classified as major hazards.

8.1.7 The importance classification of affected objects in impoundment-affected area shall be in accordance with Table 8.1.7.

Table 8.1.7 Importance classification of affected objects in impoundment-affected area

Importance	Affected object
Important	Buildings, important structures, important infrastructure
Less important	General infrastructure, cultivated land, garden plots

8.1.8 The hazard degree classification of impoundment-affected areas shall be in accordance with Table 8.1.8.

Table 8.1.8 Hazard degree classification of impoundment-affected areas

Hazard degree	Degree of damage and loss of function of affected object
High	Severe damage, loss of service function
Low	Slight damage, no loss of service function

8.2 Principles for Defining Reservoir Bank Landslide-Affected Area

8.2.1 Before impoundment, the landslide-affected areas shall be identified in one of the following cases:

1 There are landslides involving important affected objects or general infrastructure, which are predicted unstable or less stable after impoundment.

2 There are landslides involving important affected objects, whose stability coefficient cannot meet the corresponding safety control criteria after impoundment.

8.2.2 After impoundment, the new landslide-affected areas shall also be identified in one of the following cases:

1 There are landslides that have slipped into reservoir after impoundment.

2 There are landslides that are unstable or less stable and involve important affected objects and general infrastructure deprived of service function after impoundment.

8.3 Principles for Defining Reservoir Bank Deformation-Affected Area

8.3.1 Before impoundment, reservoir bank deformation-affected areas shall also be identified in one of the following cases:

1 There are deformation or accumulation bodies involving important affected objects or general infrastructure, which are predicted unstable or less stable after impoundment.

2 There are deformation or accumulation bodies involving important affected objects, whose stability coefficient is predicted not up to the corresponding safety control criteria after impoundment.

8.3.2 After impoundment, the new reservoir bank deformation-affected area shall also be identified in one of the following cases:

1 There are deformed reservoir banks involving important affected objects.

2 There are deformed reservoir banks involving general infrastructure deprived of service function.

8.4 Principles for Defining Reservoir Bank Collapse-Affected Area

8.4.1 Before impoundment, the predicted reservoir bank collapse areas involving affected objects shall be regarded as reservoir bank collapse-affected areas.

8.4.2 After impoundment, the new bank collapse areas involving affected objects shall also be regarded as reservoir bank collapse-affected areas.

8.5 Principles for Defining Reservoir Immersion-Affected Area

8.5.1 Before impoundment, the buildings, important structures, important infrastructure, general infrastructure, cultivated land and garden plots predicted with high hazard in reservoir immersion areas should be regarded as reservoir immersion-affected areas.

8.5.2 After impoundment, the new immersion areas involving important affected objects predicted with high hazard should also be regarded as reservoir immersion-affected areas.

8.6 Principles for Defining Reservoir Karst Waterlogging-Affected Area

8.6.1 Before impoundment, the predicted reservoir karst waterlogging-affected areas shall be identified in one of the following cases:

1 The predicted reservoir karst areas involving important affected objects deprived of service function where impoundment might cause or aggravate karst waterlogging should be regarded as reservoir karst waterlogging-affected areas.

2 The reservoir karst waterlogging-affected area shall be defined depending on the waterlogging time and the degree of hazard to the affected objects.

8.6.2 After impoundment, the karst waterlogging areas involving important affected objects with high hazard should also be regarded as reservoir karst waterlogging-affected areas.

8.7 Principles for Defining Reservoir Mined-Out Deformation-Affected Area

8.7.1 Before impoundment, the existing mined-out areas involving important affected objects where impoundment might cause deformation should be regarded as reservoir mined-out deformation-affected areas.

8.7.2 After impoundment, the new mined-out deformation areas involving important affected objects with high hazard should also be regarded as reservoir mined-out deformation-affected areas.

9 Defining Reservoir Bank Landslide-Affected Area

9.1 Basic Geological Conditions of Reservoir Bank Landslide Area

9.1.1 The geographic location, hydrometeorology, and geological investigation of landslide area shall be briefly described.

9.1.2 The topography and geomorphy, stratigraphy and lithology, geological structure, and hydrogeological conditions in the landslide area shall be described in detail.

9.2 Characteristics of Reservoir Bank Landslide Development

9.2.1 The distribution range, zoning, material composition, thickness, and volume of landslide mass shall be described in detail.

9.2.2 The physical and mechanical properties of landslide mass shall be described in detail.

9.2.3 The signs and characteristics of deformation and failure, genesis and development pattern of deformation, and the deformation and failure stage of landslide shall be described in detail.

9.2.4 The investigation content, methods and evaluation of reservoir bank landslide shall comply with the current standards of China GB 50287, *Code for Hydropower Engineering Geological Investigation*; and NB/T 10131, *Specification for Reservoir Area Engineering Geological Investigation of Hydropower Projects*.

9.3 Analysis and Assessment of Reservoir Bank Landslide Stability

9.3.1 The structure, failure mode, stability state, stability coefficient, stability analysis and evaluation of reservoir bank slopes shall be described in detail.

9.3.2 In the analysis and evaluation of landslide stability, the stability analysis methods, engineering slope class, stability safety factor, stability calculation and evaluation shall comply with the current sector standard DL/T 5353, *Design Specification for Slope of Hydropower and Water Conservancy Project*.

9.4 Determining Reservoir Bank Landslide-Affected Area

9.4.1 The identification of reservoir landslide-affected area shall follow the basic principles for identifying the impoundment-affected area and reservoir landslide-affected area. The landslide stability in natural state shall be analyzed.

The possibility of landslide failure during the construction period and the operation period shall be both analyzed. The potential impact on cities and towns, residential areas, cultivated land, garden plots, main traffic lines, and special facilities shall be evaluated. The range of the affected area shall be identified.

9.4.2 The identification results of landslide-affected areas shall be described in detail, and the identification result tables and maps shall be provided.

9.4.3 Suggestions on treatment, relocation and monitoring should be proposed for the landslide-affected area.

10 Defining Reservoir Bank Deformation-Affected Area

10.1 Basic Geological Conditions of Reservoir Bank Deformation Area

10.1.1 The geographic location, hydrometeorology and geological survey of reservoir bank deformation areas shall be briefly described.

10.1.2 The topography and geomorphy, stratigraphy and lithology, geological structure, and hydrogeological conditions of reservoir bank deformation areas shall be described in detail.

10.2 Characteristics of Reservoir Bank Deformation Area

10.2.1 The distribution range, zoning, material composition, thickness and volume of reservoir bank deformation areas shall be described in detail.

10.2.2 The physical and mechanical properties of rock-soil mass of reservoir bank deformation areas shall be described in detail.

10.2.3 The signs and characteristics of deformation and failure, contributing factors, genesis and development pattern of deformation, and deformation stage of reservoir bank deformation areas shall be described in detail.

10.2.4 The investigation content, methods and evaluation of reservoir bank deformation areas shall comply with the current standards of China GB 50287, *Code for Hydropower Engineering Geological Investigation*; and NB/T 10131, *Specification for Reservoir Area Engineering Geological Investigation of Hydropower Projects*.

10.3 Stability Analysis and Assessment of Reservoir Bank Deformation Area

10.3.1 The structure, stability state, stability coefficient, and stability analysis and evaluation of slopes in the reservoir bank deformation areas shall be described in detail.

10.3.2 In the stability analysis and evaluation of reservoir bank deformation areas, the stability analysis methods, engineering slope class, stability safety factor, and stability calculation and evaluation shall comply with the current sector standard DL/T 5353, *Design Specification for Slope of Hydropower and Water Conservancy Project*.

10.4 Determining Reservoir Bank Deformation-Affected Area

10.4.1 The identification of reservoir bank deformation-affected area shall follow the basic principles for identifying impoundment-affected areas and reservoir bank deformation-affected areas. Based on the predicted impact range of reservoir bank deformation, the potential impact on and hazards to cities and towns, residential areas, cultivated land, main traffic lines, and special facilities shall be evaluated.

10.4.2 The zoning results of reservoir bank deformation-affected areas shall be described in detail, and the identification result tables and maps shall be provided.

10.4.3 Suggestions on treatment, relocation and monitoring should be proposed for the reservoir bank deformation-affected area.

11 Defining Reservoir Bank Collapse-Affected Area

11.1 Basic Geological Conditions of Reservoir Bank Collapse Area

11.1.1 The geographic location, hydrometeorology and investigation of reservoir bank collapse areas shall be briefly described.

11.1.2 The topography and geomorphy, stratigraphy and lithology, geological structure and hydrogeological conditions of reservoir bank collapse areas shall be described in detail.

11.2 Analysis on Causes of Reservoir Bank Collapse

11.2.1 The parameters of bank slope soil mass for collapse prediction shall be specified. In prediction calculation, the stable slope angle of each bank section shall be determined based on the test results and investigation data.

11.2.2 The hydrodynamic environment conditions, signs of deformation and failure, structural type of reservoir bank, contributing factors, bank collapse mode, bank collapse mechanism, development pattern, and stages of deformation and destruction of the reservoir banks shall be described in detail.

11.2.3 The investigation content, methods and evaluation of bank collapse shall comply with the current standards of China GB 50287, *Code for Hydropower Engineering Geological Investigation*; and NB/T 10131, *Specification for Reservoir Area Engineering Geological Investigation of Hydropower Projects*.

11.3 Prediction and Evaluation of Reservoir Bank Collapse

11.3.1 The final width of reservoir bank collapse areas after impoundment shall be predicted, the hazard of bank collapse shall be evaluated, and the range of reservoir bank collapse-affected area shall be determined.

11.3.2 Qualitative and quantitative methods should be used for reservoir bank collapse prediction, including engineering geological analogy method, graphic method and calculation method.

11.3.3 The stability analysis and assessment of sliding bank collapse shall comply with the current sector standard DL/T 5353, *Design Specification for Slope of Hydropower and Water Conservancy Project*.

11.4 Determining Reservoir Bank Collapse-Affected Area

11.4.1 The identification of reservoir bank collapse-affected areas shall follow the basic principles for defining impoundment-affected areas and reservoir bank collapse-affected areas. According to the predicted impact degree of bank collapse under different reservoir water levels, the possible impacts on and hazards to cities and towns, residential areas, cultivated land, and garden plots, main traffic lines and special facilities shall be evaluated.

11.4.2 The zoning results of reservoir bank collapse-affected areas shall be described in detail, and the identification result tables and maps shall be provided.

11.4.3 Suggestions on treatment, relocation and patrol inspection of bank collapse-affected areas should be put forward.

12 Defining Reservoir Immersion-Affected Area

12.1 Basic Geological Conditions of Reservoir Immersion Area

12.1.1 The genetic type, structure, material composition, range, ground elevation, and topographic gradient of the river valley terraces and other accumulation platforms in the reservoir immersion area shall be described in detail.

12.1.2 The hydrogeological conditions of the reservoir immersion area, the type of groundwater, and the outcropping and recharge, runoff and discharge conditions of groundwater shall be described in detail.

12.2 Analysis on Causes of Reservoir Immersion

12.2.1 The types of main crops, root layer thickness, history and current situation of soil salinization and swamping in the reservoir immersion area shall be described in detail.

12.2.2 The foundation type and burial depth of buildings in the reservoir immersion area, deformation and destruction of houses, and degree of impact on the human settlement environment shall be described in detail.

12.2.3 The distribution, elevation, structure and morphological characteristics of mines and underground structures around the reservoir immersion area shall be described in detail.

12.3 Prediction and Evaluation of Reservoir Immersion

12.3.1 The critical groundwater burial depth, the predicted immersion scope, type and scale, and the hazard evaluation shall be described in detail.

12.3.2 Prediction and evaluation of reservoir immersion impacts shall comply with the current national standard GB 50287, *Code for Hydropower Engineering Geological Investigation*.

12.4 Determining Reservoir Immersion-Affected Area

12.4.1 The identification of reservoir immersion-affected areas shall follow the basic principles for defining impoundment-affected areas and reservoir immersion-affected areas. According to the predicted range of reservoir immersion-affected areas, the potential impacts on and hazards to cities and towns, residential areas, cultivated land, and garden plots, main traffic lines and special facilities shall be evaluated.

12.4.2 The zoning results of reservoir immersion-affected areas shall be described in detail, and the identification result tables and maps shall be provided.

12.4.3 Suggestions on treatment, relocation and monitoring of reservoir immersion-affected areas should be put forward.

13 Defining Reservoir Karst Waterlogging-Affected Area

13.0.1 The topography and geomorphy, stratigraphy and lithology, geological structure, karst development characteristics and karst hydrogeological conditions in the karst waterlogging area shall be described in detail.

13.0.2 The distance, elevation difference and groundwater gradient from the basin, depression and valley to the riverbed and reservoir, the distribution, plane and profile morphology, scale and development history of underground karst river system, the location and elevation of the sinkhole entrance and outlet of underground river, and the existence and cause of natural karst waterlogging shall be described in detail.

13.0.3 Conditions for formation of karst waterlogging areas shall include the following:

1 The characteristics of karst conduits and the relationship between the normal pool level and the elevation of the underground river outlet in reservoir shall be described in detail .

2 The relationship between the elevation of karst basin, depression or valley and the normal pool level shall be described in detail.

13.0.4 The prediction and evaluation of karst waterlogging-affected areas shall meet the following requirements:

1 The impact of impoundment on the discharge of underground river due to outlet submerging, on the discharge of karst basin, depression and valley, and on the drainage of karst basin, depression and valley caused by local silting of underground karst conduit shall be described in detail.

2 The prediction method and parameters of karst waterlogging shall be described in detail, and backwater height shall be predicted.

3 The range and scale of karst waterlogging shall be described in detail.

13.0.5 The identification of karst waterlogging-affected areas shall comply with the following provisions:

1 The identification of karst waterlogging-affected areas shall follow the basic principles for defining impoundment-affected areas and reservoir karst waterlogging-affected areas. According to the predicted impact range of karst waterlogging, the impacts on and hazards to buildings, structures, infrastructure, and crop types shall be evaluated.

2 The identification results of karst waterlogging-affected areas shall be described in detail, and the identification result tables and maps shall be provided.

3 Suggestions on treatment, relocation and monitoring of karst waterlogging-affected areas should be put forward.

13.0.6 For new karst waterlogging areas after impoundment, data collection, causes analysis and hazards evaluation shall be conducted, and suggestions on treatment and monitoring shall be put forward.

14 Defining Reservoir Mined-Out Deformation-Affected Area

14.0.1 The topography and geomorphy, stratigraphy and lithology, ore beds, geological structure and hydrogeological conditions of reservoir mined-out areas shall be described in detail.

14.0.2 The analysis of formation conditions of mined-out deformation shall include the following:

1 The position, shape, size, depth and extension direction of surface subsidence pits and cracks of mined-out deformation and their relationship with stratigraphy and geological structure in mined-out areas shall be described in detail.

2 The distribution range of mined-out deformation, mining start date and end date, mining depth, thickness and method, position and size of the main tunnel, collapse, support, backfilling and water filling of main tunnel shall be described in detail.

3 The relationship between the elevation of mined-out deformation area and the normal pool level shall be described.

14.0.3 The impact prediction of mined-out deformation should meet the following requirements:

1 The characteristic values of surface displacement and deformation in the mined-out deformation area should be described in detail.

2 The prediction methods for mined-out deformation-affected areas may be the critical burial depth method, numerical analysis method, etc.

3 The range of mined-out deformation areas affected by impoundment and drawdown should be predicted.

14.0.4 The identification of mined-out deformation-affected areas shall meet the following requirements:

1 The identification of mined-out deformation-affected areas shall follow the basic principles for defining impoundment-affected areas and mined-out deformation-affected areas. Based on the predicted impact range of mined-out deformation, the impacts on and hazards to buildings, structures, infrastructure, cultivated land, and garden plots shall be evaluated.

2 The zoning results of mined-out deformation-affected areas shall be described in detail, and the identification result tables and maps shall be provided.

3 Suggestions on treatment, relocation and monitoring of mined-out deformation-affected areas should be put forward.

15 Geological Results for Defining Impoundment-Affected Area

15.0.1 The geological results shall be summarized for the identification of impoundment-affected areas.

15.0.2 The summary of geological results may include the situation, type, characteristics, and range of the affected areas, affected objects, degree of hazards, and suggestions on treatment methods.

16 Conclusions and Suggestions

16.0.1 The regional geology and seismicity, basic geological conditions of reservoir area, engineering geological segmentation and evaluation of reservoir bank, conclusions and suggestions should be summarized.

16.0.2 For different impoundment-affected areas, the conclusions and suggestions on the identification of reservoir landslide-affected area, reservoir deformation-affected area, reservoir bank collapse-affected area, reservoir immersion-affected area, reservoir karst waterlogging-affected area and reservoir mind-out deformation-affected area should be summarized, respectively.

16.0.3 The report should indicate the geological risks in corresponding impoundment-affected areas and give suggestions on future work based on the conclusions of impoundment-affected area identification.

16.0.4 Suggestions on emergency measures of impoundment-affected areas should be put forward considering the actual project conditions after impoundment.

Appendix A Statistics of Geological Results for Defining Impoundment-Affected Areas

Table A Statistics of geological results for defining impoundment-affected areas

Type of affected area	No.	Conditions of affected area					Characteristics of affected area	Range of affected area			Affected object	Degree of hazard	Suggestion	Remarks
		Coordinates	Name	Location	Bank (left/right)	Distance from dam (km)		Elevation of front edge (m)	Elevation of rear edge (m)	Area (m^2)				
	1													
	2													
	…													

NOTES:

1 The types of impoundment-affected areas include the areas affected by landslide, bank deformation, bank collapse, immersion, and karst waterlogging, mined-out deformation, and reservoir leakage.

2 For bank collapse-affected areas, the front and rear edge elevations refer to those corresponding to the predicted width of final collapse.

3 The affected objects are residential areas, cultivated land, garden plots, houses, buildings, structures, infrastructure, special facilities, etc.

4 The starting point for area calculation refers to the top elevation of reservoir inundation area.

5 Suggestion refers to the suggestion on treatment of impoundment-affected areas.

Appendix B Contents of Special Geological Report on Defining Impoundment-Affected Areas

1 Overview

2 Regional Geology and Seismicity

3 Basic Geological Conditions of Reservoir

3.1 Topography and Geomorphy

3.2 Stratigraphy and Lithology

3.3 Geological Structure of Reservoir Area

3.4 Geophysical Phenomena

3.5 Hydrogeological Conditions

4 Engineering Geology Segmentation and Evaluation of Reservoir Bank

4.1 Segmentation Principles of Reservoir Bank

4.2 Segmentation and Assessment of Reservoir Bank

5 Geological Principles for Defining Impoundment-Affected Area

6 Defining Reservoir Bank Landslide-Affected Areas

6.1 Basic Geological Conditions of Reservoir Landslide Area

6.2 Characteristics of Reservoir Bank Landslide Development

6.3 Analysis and Assessment of Reservoir Bank Landslide Stability

6.4 Determining Reservoir Bank Landslide-Affected Area

7 Defining Reservoir Bank Deformation-Affected Area

7.1 Basic Geological Conditions of Reservoir Bank Deformation Area

7.2 Characteristics of Reservoir Bank Deformation Area

7.3 Stability Analysis and Assessment of Reservoir Bank Deformation Area

7.4 Determining Reservoir Bank Deformation-Affected Area

8 Defining Reservoir Bank Collapse-Affected Area

8.1 Basic Geological Conditions of Reservoir Bank Collapse Area

8.2 Analysis of Formation Conditions of Reservoir Bank Collapse

8.3 Prediction and Assessment of Reservoir Bank Collapse

8.4 Determining Reservoir Bank Collapse-Affected Area

9 Defining Reservoir Immersion-Affected Area

9.1 Basic Geological Conditions of Reservoir Immersion

9.2 Analysis of Formation Conditions of Reservoir Immersion

9.3 Prediction and Evaluation of Reservoir Immersion

9.4 Determining Reservoir Immersion-Affected Area

10 Defining Reservoir Karst Waterlogging-Affected Area

11 Defining Reservoir Mined-Out Deformation-Affected Area

12 Geological Results for Defining Impoundment-Affected Area

13 Conclusions and Suggestions

14 Attachments

(1) Stage Report Review and Relevant Approval Opinions

(2) Review Comments on Special Reports

15 Attached Drawings

(1) Comprehensive Geological Plan of Reservoir Area

(2) Distribution Plan of Impoundment-Affected Area

(3) Geological Plan and Profile of Typical Impoundment-Affected Area

Explanation of Wording in This Specification

1 Words used for different degrees of strictness are explained as follows in order to mark the differences in executing the requirements in this specification.

1) Words denoting a very strict or mandatory requirement:

"Must" is used for affirmation; "must not" for negation.

2) Words denoting a strict requirement under normal conditions:

"Shall" is used for affirmation; "shall not" for negation.

3) Words denoting a permission of a slight choice or an indication of the most suitable choice when conditions permit:

"Should" is used for affirmation; "should not" for negation.

4) "May" is used to express the option available, sometimes with the conditional permit.

2 "Shall meet the requirements of..." or "shall comply with..." is used in this specification to indicate that it is necessary to comply with the requirements stipulated in other relative standards and codes.

List of Quoted Standards

GB 50287, *Code for Hydropower Engineering Geological Investigation*

NB/T 10131, *Specification for Reservoir Area Engineering Geological Investigation of Hydropower Projects*

DL/T 5353, *Design Specification for Slope of Hydropower and Water Conservancy Project*

DL/T 5376, *Specification of Scoping for Land Acquisition of Hydroelectric Projects*